Russia's Coastal Missile Shield Bal-E & Bastion

Mobile Coastal Cruise Missile Complexes

HUGH HARKINS

Copyright © 2021 Hugh Harkins, FRAS, MIstP, MRAeS

All rights reserved.

ISBN: 1729845045
ISBN-13: 978-1729845042

Russia's Coastal Missile Shield
Bal-E & Bastion

Mobile Coastal Cruise Missile Complexes

© Hugh Harkins 2021

Createspace Independent Publishing Platform
United States

ISBN 10: 1729845045
ISBN 13: 978-179845042

This volume edition first published in 2021
Previous volume edition published in 2018

The Author is identified as the copyright holder of this work under sections 77 and 78 of the Copyright Designs and Patents Act 1988

Cover design © Centurion Publishing and Createspace Independent Publishing Platform

Page layout, concept and design © Centurion Publishing

All rights reserved. No part of this publication may be reproduced, stored in a retrieval system, transmitted in any form, or by any means, electronic, mechanical or photocopied, recorded or otherwise, without the written permission of the publisher

The publisher and author would like to thank all organisations and services for their assistance and contributions in the preparation of this volume: CDB-Titan, Concern Morinformsystem-Agat, JSC Avanguard, JSC Concern Granit-Electron, JSC MIC Mashinostroyenia (Joint Stock Company Military Industrial Corporation-Scientific and Production Machine Building Association), JSC NPP Radar MMS, JSC Typhoon, Matra British Aerospace Dynamics Alenia, Ministry of Defence of the Russian Federation, NPO Mars, OJSC Dubna Machine Building Plant, PA Strela, Rosoboronexport, Tactical Missiles Corporation, Volat

Citation guide: Titan (CDB-Titan), Morinformsystem-Agat (Concern Morinformsystem-Agat), Avanguard (JSC Avanguard), Granit-Electron (JSC Concern Granit-Electron), Mashinostroyenia (JSC MIC Mashinostroyenia), Radar MMS (JSC NPP Radar MMS), Typhoon (JSC Typhoon), MBDA (Matra British Aerospace Dynamics Alenia), MODRF (Ministry of Defence of the Russian Federation), Mars (NPO Mars), BBM (OJSC Dubna Machine Building Plant), Strela (PA Strela), Rosoboronexport (Rosoboronexport), TMC (Tactical Missiles Corporation), Volat (Volat)

CONTENTS

	INTRODUCTION	vii
1	SOVIET COASTAL CRUISE MISISLE LEGACY OVERVIEW	1
2	BAL-E	5
3	BASTION	29
4	OPERATIONAL	51
5	GLOSSARY	59

INTRODUCTION

The Bal-E and Bastion mobile coastal missile complexes, introduced to service with the Russian Federation in the second decade of the twenty first century, are new generation weapon systems designed to deny access to a potential enemy – in particular the NATO (North Atlantic Treaty Organisation) alliance, which is perceived as the main threat to the Russian Federation – to Russia's vast coastlines, sea lines of communications, ports and naval bases. Bal-E and Bastion capabilities are complementary to each other and form two of the three surface based medium range and inner elements of Russia's overall defence in depth policy to counter NATO land attack cruise missile armed warships, aircraft carrier battle groups and amphibious assault groups threatening Russia's territorial integrity.

This volume sets out to detail the Bal-E and Bastion complexes in domestic Russian service and extends to the stationary Bastion complex that has been developed and tested, but, as of 2018, remains un-fielded. All technical data relating to the respective weapon systems and their components have been provided by the respective design bureaus and manufacturers, as has much of the imagery and graphics, with additional impute from the MODRF (Ministry of Defense of the Russian Federation). Whilst the use of an 'E' at the end of a specific Russian Federation weapon system designation often denotes an export standard, the use of the Bal-E designation is employed by the MODRF in reference to the domestic standard.

This revised volume replaces the volume published in 2018. It is generally identical except in respect to the calculation of Bastion complex missile flight time: The value for the entire flight duration is included, whereas the 2018 volume carried only the value for the terminal flight phase.

1

SOVIET COASTAL CRUISE MISSILE LEGACY OVERVIEW

Cruise missiles were a cornerstone of Soviet defence planning from the early 1950's through the dissolution of the Soviet Union on 25 December 1991 and remain a major element of the defence planning of the Soviet Union's major successor state, the Russian Federation, in the second decade of the twenty first century. A number of early programs resulted in cruise missiles for air launch, submarine launch, surface warship launch and shore based launch. The early Soviet shore based cruise missile systems were the forerunners of the advanced Bal-E and Bastion complexes in service with the Russian Federation in the twenty first century.

The first generation Soviet shore-based ASCM (Anti-Ship Cruise Missile) complex KSS (C-2 (P-2) 'Sopka', was developed from the A. Mikoyan OKB air launched KS ASCM, which, in 1951, entered production and would later arm the Tupolev Tu-16KS and Tu-16K (entered production in 1959) four jet missile carriers. The Tu-16K-11-16, which constituted an element of the outer layer of ASCM defence, had a maximum speed of 1050 km/h, an operating ceiling of 15000 m, an operational range of 7800 km and could carry two KSR-2 and, later, KSR-5 (Tu-16K-26/16-10-26) or KSR-11) ASCM. In order to counter threats that had eluded the outer defensive layer of air launched ASCM, the C-2 shore launched ASCM was developed under the design leadership of A.I. Mikoyan with designers A. Ya. Bereznyak, A.A. Lazarev and M.I. Gurevivh. The outer lines of the missile were externally identical to the KS. The wings of the C-2, which employed radio+ARRLSN for navigation/guidance, were folded for storage and extended for cruise flight. The missile was launched under the power of an SPRD-15 engine before entering cruise flight under the power of a single RD-500K turbojet engine. Production of the C-2 commenced in 1954 and the complex was adopted for service with the Soviet Union in 1957. In the same timeframe the CS-7 (KS-7) FKK-1 surface launched radio-beam guided cruise missile was developed by Mikoyan as the armament of the FKR-1 ground mobile complex for Soviet ground forces. This missile complex, which adopted a development of the KSS airframe and the same propulsion systems, entered production in 1954 (DBM).

The experience gained in the development and fielding of the early ground based cruise missile complexes paved the way for the introduction of more advanced systems – Redut (Redout) and Rubezh that served the Soviet Union through the latter part of the Cold War, the break-up of the Soviet Union in December 1991 and into the second decade of the twenty first century when they began to be replaced by Bal-E and Bastion.

A Kh-35E missile is launched from a Bal-E self-propelled launch vehicle of the Russian Federation Pacific Fleet in July 2018. MODRF

OKB-52 - Experimental Design Bureau-52 (now JSC MIC Mashinostroyenia) was also involved in cruise missile development, building on research conducted in the period 1944-1953. In 1954, the same year that the C-2 missile complex was ready for production, a 'Special Design Group' stood up at Factory No.500 located in Tushino. The main area of study for this group at that time was development of the 10 XH air launched cruise missile (flying bomb), but the organisation would evolve into the realms of submarine, surface ship and land based cruise missile design. The 'Special Design Group' was, in 1955, reorganised to form the USSR (Union of Soviet Socialist Republics) Experimental Design Bureau No.52 and relocated to Reutov, Moscow region (Mashinostroyenia). Among the programs this bureau was responsible for was the P-5 and P-5D seaborne ASCM complexes and the C-5 land based complex, which was designed to strike shore and inland targets (Mashinostroyenia). The P-5 variants & the P-6 constituted the main armament on Soviet submarines of the projects 644, 651, 665, 675 and 644-D (P-5D). The P-35 complex was developed as the ASCM armament for a number of Soviet surface warships. A land based variant of the P-35 was developed to arm the mobile Redut complex – a single P-35 missile was carried on a ZIL-135K 8x8 wheeled chassis (Mashinostroyenia).

Redut (Redout) Coastal Missile System – data furnished by MODRF

Mass of launcher and missile: 21 tons
Starting weight of missile: 4500 kg
Maximum flight range: 270 km
Flight altitude range: 40/4000/7000 m
Maximum flight speed: Mach 1.5
PU [transport vehicle] Maximum speed: 40 km/h
Cruise range: 500 km
Crew: 5 persons

Rubezh shore based ASCM complex. MODRF

The ICB Rainbow developed C-3 (Rubezh shore based ASCM complex, which entered production in 1975, was developed as a mobile system based on a MZKT MAZ-543M 8x8 wheeled chassis. The MAZ-543M, with a load capacity of 22200 kg, and engine power of 575 PS, could carry two P-15M/21 ASCM missiles at a maximum road speed of 60 km/h. The MZKT-543M chassis also forms the basis of the Bereg mobile coastal artillery complex (Volat). This complex, which is armed with an ACS 130 mm calibre gun, can detect targets at ranges out to 35 km and effectively strike targets travelling at speeds of up to 200 knots at ranges out to 20 km with an extended maximum range of 23 km, although accuracy would drop off. Maximum rate of fire is 12-14 rounds per minute and a total ammunition load of 44 high explosive rounds is carried (Rosoboronexport).

In the immediate post-Cold War years of the 1990's and through the first decade of the twenty first century the Russian Federation was reliant on the 1970's era Redut and Rubezh CMS/ASCM complexes, which were still formidable weapon systems, if somewhat outdated. Going into 2018, Redut and Rubezh complexes were still nominally on strength with Russian Federation coastal missiles forces – the former certainly remaining in service with the Pacific Fleet coastal troops (MODRF). Both complexes are expected to be retired as Bal-E and Bastion are firmly established in service, although no out of service date has been announced for either weapon system.

2

BAL-E

The Bal-E ASCM (Anti-Ship Cruise Missile) complex in service with the Russian Federation is designed to engage a diversity of target sets, including surface warships of all sizes, auxiliary vessels and amphibious assault ship at ranges out to 130 km when armed with the Kh-35E (3M-24E) missile. Range is extended to 260 km when armed with the Kh-35UE missile (while the Kh-35UE is an option for both domestic and potential export operators of Bal-E there is, in 2018, no confirmation of this variant being in operation with the various fleets of the Russian Federation coastal missile forces). The Bal-E complex is tasked with defending Russian Federation naval facilities, various costal installations, territorial waters, offshore sea lines of communications and, in time of open hostilities, provide access denial to enemy shipping (including warships) out to the ranges stated above (TMC).

A typical Bal-E unit consists of up to four 3S-60E self-propelled LV (Launch Vehicles) – MZKT-7930 chassis – up to four 3F-60E self-propelled TLV (Transport re-Loader Vehicles) – MZKT-7930 chassis – and up to two 3Ts-61E self-propelled CCC (C3 (Command, Control & Communications)) vehicles (TMC, Morinformsystem-Agat & Rosoboronexport). Each LV can be armed with eight Kh-35E ASCM and can be reloaded with a full complement of eight Kh-35E carried on the TLV. The number of self-propelled TLV required is dependent on the number of LV in a specific unit (Morinformsystem-Agat) – four 3S-60E self-propelled LV would be accompanied by four self-propelled 3F-60E TLV, each of which can accomplish a full eight missile reload for a LV in a time of 30-40 minutes (TMC). The CCC centre facilitates centralised control over all self-propelled LV in a specific unit, providing designation of targets, which are allocated to the most optimum launch platform(s). All of the LV, TLV, and CCC vehicles are equipped with integral stand-alone power sources and can operate from external power sources and the entire CMS (Coastal Missile System) unit is designed to operate in conditions of NBC (Nuclear Biological and Chemical) contamination of the surface or airborne environment (Typhoon).

A MZKT-7930-300 chassis is put through its paces during development trials. Volat

The various mobile elements of a Bal-E unit are based on high mobility rough terrain vehicles. The Volat MZKT-7930 was selected as the chassis for the major elements of the Bal-E and Bastion mobile CMS.

MZKT-7930 performance characteristics – data furnished by Volat

Wheel drive: 8x8
Max speed: 70 km/h
Gradeability: 55%
Angle of approach: 25°
Angle of departure: 35°
Ground clearance: 400 mm
Climbing steps: up to 350 mm
Side slope: 40%
Turning radius (curb to curb): 15 m
Fording depth with pas: 1.4 m
Trench width: 2 m
Cruising range: 1000 km
Operating temperature: -45° to +50° C
Fuel tank capacity: 2 x 385 litre
Air transportable: can be carried in Antonov An-22 'Cock' and An-124 'Ruslan' military transport aircraft in service with the Russian Federation Aerospace Forces

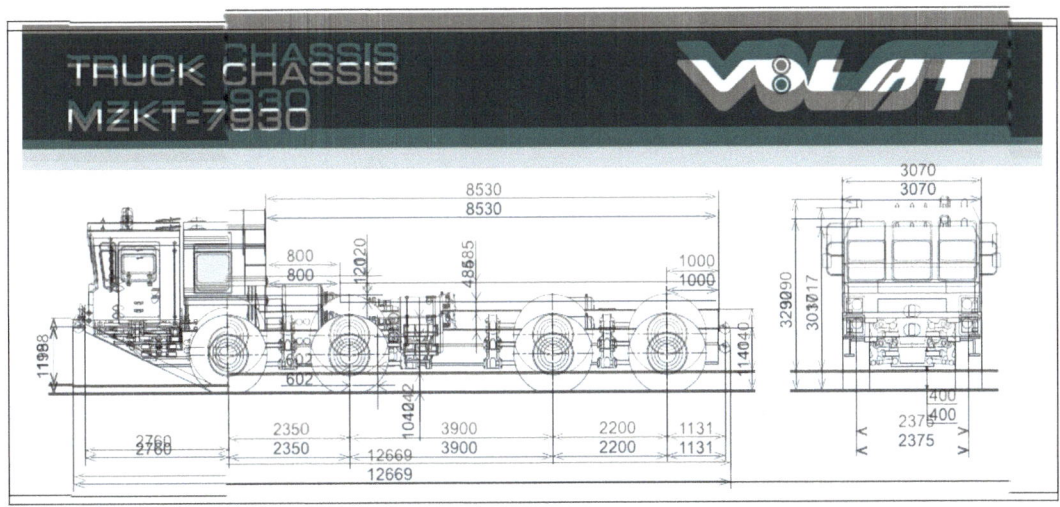

Graphic depicting the overall dimensions of the MZKT-7930 chassis used in the Bal-E and Bastion coastal missile system complexes - values are given in mm. Volat

MZKT-7930 characteristics - data furnished by Volat

Weights
- Gross vehicle weight: 45200 kg
- Curb Vehicle weight: 20500 kg
- Payload: 24200 kg
- Permissible weight on axle
- 1st and 2nd axles: 2 x 11050 kg
- 3rd and 4th axles: 2 x 11550 kg

Engine & Transmission
- Type: YAMZ-846 diesel V-shape, 8-cylinder
- Power: 370 kW (503 h.p.) at 2100 min^{-1}
- Maximum torque: 1960 Nm (200 kg/cm)
- Transmission type: YAMZ-202-04; manual
- Number of gears: 9 forward and 1 reverse
- Transfer case: 2-speed with lockable inter-bogie differential
- Gear ratio: 1:1.00/1:1.601
- Driving axles: central reduction gears and inter-axle and cross-axle differentials lock, planetary wheel hub reduction.
- First & second axles: steerable, driven
- Suspension: independent torsional wheel suspension, heavy-duty shock absorbers
- Steering system: left/right hand drive, hydraulic power assisted
- Brake system: duel-circuit with pneumo-hydraulic drive
- Tyres: 1500x600-635, CTIS
- Frame: low torsion ladder frame, coupled with bolted cross-members, steel bumper. Frame equipped in front with two towing forks and rear towing hook

MZKT-7930 characteristics - data furnished by Volat (continued from page 8)

Cab: fibre-glass, 2 door, 3 seat, adjustable steering wheel, lighting instrument panel, interior lighter and reading lamp
Electrical Equipment
Nominal voltage: 24 v
Batteries: 4 pcs, 380 Ah
Single-wire, screened
Options: include a cable winch for recovery of other vehicles or self-recovery; Powerful heater/ventilator; air conditioner; auxiliary heater; 2 large central LCD [liquid crystal display]; Run flat types and an armoured cab Note: It is unclear which, if any, of the options are incorporated in the MZKT-7930 operated by the Bal-E and Bastion CMS in service with the Russian Federation

Bal-E coastal missile system complex self-propelled launch vehicle in the mobile ready configuration (top) and being put into the launch configuration (above). Typhoon & MODRF

Bal-E characteristics – data furnished by TMC, MODRF, Concern Morinformsystem-Agat, JSC Typhoon and Rosoboronexport

Surface detection range of the Monolit-B active radar channel
   With the antenna positioned 12 m above sea level: 35 km
   In sea surface duct: 100 km
   In super refraction: up to 250 km
Maximum number of tracked targets
   By active radar: 30
   By passive radar in detection mode: 50
   By passive radar in targeting mode: 10
Monolit-B passive radar channel surface target detection range: up to 450 km
Missile engagement range parameters (Kh-35E): 5-130 km
Maximum number of targets simultaneously engaged by one full salvo: 24 [taken to be three launch vehicles each with a complement of eight Kh-35E missiles]
Deployment time after march: no more than 15 minutes
Missile number: 64 (8 in each of 4 LV and 8 in each of 4 TLV)
Time between missile launches on a single launcher: ~3 seconds
Maximum elevation of fire position above sea level: up to 1000 m
Distance of launcher form coastline: up to 10 km
Launch platform: MZKT-7930 cross-country four axle wheeled chassis
Maximum speed: 60 km/h on road surface and 20 km/h off-road surface
Unrefueled range: 850 km [Bal-E complex specific]
Missile launch weight: ~620 kg
Crew: 11

Previous page: The 3S-60E transport re-loader vehicle of the Bal-E coastal missile complex. Each transport re-loader vehicle accommodates a load of eight Kh-35E anti-ship cruise missiles and is outfitted with the necessary lifting equipment to facilitate removal of the empty launch containers from the launch vehicles and replacement with the new missile load in their self-contained launch containers. This page: The command, control and communications vehicle of the Bal-E complex is centered on the Monolit-B surveillance and reconnaissance system that can also operate as a stand-alone surveillance system, providing targeting data for other systems such as the Bastion coastal missile system. Morinformsystem-Agat

Bal-E launch vehicle with Kh-35E missile (top) and a MZKT-7930 cab section with missile blast screens deployed (above). Rosoboronexport/MODRF

The Bal-E launch vehicle can receive target designation data from the command vehicles acquired data or data from higher echelon surveillance and command posts (Morinformsystem-Agat). The command vehicle provides for high-speed reception of various data sets handed down from upper level command elements and various tactical and strategic reconnaissance systems, such as aircraft and orbiting satellites. The command and control of the complex is conducted through secure digital automated communications, including transmissions of data, encrypted or unencrypted (TMC). This allows for very fast data transfer to and from higher command assets, reconnaissance/surveillance assets and unit command posts. At the heart of the Monolit-B command vehicle is the jam resistant Mineral-E radar complex, which operates in active and passive radar channels and can function in a dense ECM (Electronic Countermeasures) environment to detect and select targets against backgrounds of not only benign, but active and passive interference, and provide target classification and tracking. When operating in the passive mode two Mineral-E complexes on separate vehicles are used in order to provide triangulation for target detection (Morinformsystem-Agat & Typhoon). The Monolit-B complex provides radar coverage of air and surface objects at over the horizon ranges. The system can detect and track surface targets by employment of radiolocation signals in both passive and active modes and surface target IFF (Identification Friend or Foe) interrogation. The system provides tracking and classification of all types of surface borne targets and low-speed, low altitude airborne targets (Typhoon).

Graphic depicting the Monolit-B element of the Bal-E complex. Rosoboronexport

· The command, control and communications units provide target designation and facilitate assigning missile launchers to the most appropriate target (Morinformsystem-Agat) and, as noted above, a system of operational control provides for handing off targets to the most optimal launch platform. For example the most appropriately positioned launcher may not be at readiness, therefore, the target would be assigned to another launcher. The different fire control options include the ability for a single mobile command and control complex to oversee the targeting of a single missile launch or a salvo launch from a single Bal-E group or for two such command and control complexes to oversee the operation of two such group salvoes with each launcher independent of the other (Rosoboronexport). Combat effectiveness is enhanced through the ability to conduct single or salvo launch in stand-alone mode, whereby target designation is handed down from higher echelon stations or another Bal-E unit linked through centralised command and control infrastructure. A salvo launch can consist of up to eight missiles from a single launch platform, several missiles from several launch platforms or all 32 missiles carried by the four launch vehicles in order to swamp an enemy task force or amphibious landing force or, indeed, an auxiliary force group of ships (Morinformsystem-Agat & TMC). Ministry of Defence of the Russian Federation data suggests that a Bal-E unit with four LV can launch 32 missiles in salvo with a three second interval between launches on each respective LV (MODRF). Following a 32 missile salvo the four LV grouping is designed to be ready to launch a further 32 missile salvo within 30-40 minutes (TMC).

Page 13: Monolit-B of the Bal-E command, control and communications vehicle in the deployed configuration. Page 14: Two photographs of the Bal-E in the deployed configuration with the Mineral-E radar antenna in various stages of rotation. This page: A Monolit-B complex (background) operating as an integral part of a Bal-E coastal missile system (launch vehicle in foreground). Rosoboronexport/MODRF

Monolit-B can operate as a non-integral element of the coastal missile system. In such independent surveillance and targeting operations the system can oversee the operation of a coastal missile system and receive inputs of surveillance data from its own onboard sensors as well as a host of external sensor sources – shipborne, coastal and airborne (Typhoon). In the non-integral coastal missile system role two Monolit-B complexes would be employed, one of which would be employed for central location surveillance and targeting and the other would be responsible for peripheral location surveillance and targeting – all data being transmitted to other assets through a number of digital formats. Standard operational doctrine would have the two Monolit-B vehicles located around 30 km apart. This would allow target designation functions to be conducted in stealthy radio silent mode by means of triangulation (Typhoon). The two vehicles would exchange data on the long-range surface/air picture through radio relay. When operating in the passive channel the systems would utilise "a low-energy active radar channel", which facilitates stealthy, difficult to detect, operation due to the very low power radio wave emissions (Typhoon).

Although it can be deployed in several other functions, such as general sea surveillance and air surveillance against low altitude objects, supporting the mobile coastal missile systems is the Monolit-B's primary function. The system can input data directly into the ASCM – Kh-35E (as an element of the Bal-E complex or operating independently) or Yakhont – control system of the Bastion coastal missile complex when operating on a stand-alone basis or allocated to support a particular Bastion battery (Typhoon).

> **Monolit-B characteristics – data furnished by JSC Typhoon & Rosoboronexport**
>
> **Range of sea-surface target detection with ARdr: [active radar] 35 km under normal propagation conditions and antenna at a height of 9-12 m above sea level:** 90-100 km with driven waveguide available; up to 250 km under super-refraction conditions
> **Range of sea surface targets detection with PRdr [passive radar]:** up to 450 km
> **Maximum coverage zone of the DEORdr:** 0.2-30 km
> **Maximum number of tracked targets with ARdr:** 30
> **Maximum number of tracked targets with PRdr in detection mode:** 50
> **Maximum number of tracked targets with PRdr targeting in data generation mode:** 10
> **Maximum number of targets processed by DEORdr:** 200

There are a number of external surveillance, targeting and command, control and communications assets that are available to Bal-E and Bastion complexes. While such assets would differ for potential export operators depending on specific requirements they could include various complexes of the stationary Podsolnukh-E sea surveillance radar, the MR-10M1E (stationary) and Mys-M1E (trailer mounted mobile) coastal radar complex that can detect and track surface warships at ranges out to several hundred kilometers and pass the data on to third party assets such as a Bal-E complex. The 83t611-E CMOP (Coastal Modular Operational Post) is effectively a complex for facilitating data exchange between all of the shore based, airborne and seaborne assets operating within the control system of the relative operational area (Rosoboronexport).

> **83t611-E characteristics – data furnished by Rosoboronexport**
>
> Establishing data exchange networks based on four-wire unswitched: voice-band, telegraph, pulse and fiber optic channels; implementation of TCP/IP and X-25 data exchange; algorithms and protocols and special message protocols and formats
> The following data rates are provided: voice-band channels, 1200 bit/s; telegraph channels, 50, 100, 200 bit/s; pulse channels, 1.2, 2.4, 9.6, 16, 32, 48 and 64 kbit/s; digital channels: 2048 kbit/s
> Data exchange is provided in the following directions: shore to shore; shore to surface ship (in port or at sea); surface ship to surface ship; surface to aircraft
> The following data rates are provided: HF radio links using a special HF modem, 1200, 2400, 4800 bit/s; HF telegraph radio links in frequency telegraphy and two-channel frequency telegraphy mode, 4, 50, 75, 100, 150, 200, 300 bit/s; VHF radio links using a broadband channel, 2.4, 4.8, 9.6, 16, 32, 48 bit/s; VHF radio links using a telegraph channel, 50, 75, 100, 150, 200, 300, 600, 1200, 2400, 4800 bit/s

Russian language graphic depicting a Bal-E CMS and other assets under the overall control of an 83t611-E Coastal Modular Operational Post. English translation of key text: надводные корабли (surface ships); бпла (BPLA); ЛЕТАТЕЛЬНЫЕ АППАРАТЫ (Aircraft in flight); ПОДВОДНЫЕ ЛОДКИ (Underwater [submerged] submarine); УКВ (VHF); БРК (BRK); Цифровые радиорелейные линии (Digital radio relay lines); Радиорелейная система связи Заказчика (customers radio relay communication system); СРЕДСТВА МОНИТОРИНГА (means of monitoring); КУЗОВ-КОНТЕЙНЕР (KUZOV (body) Container); ПУЛЬТ УПРАВЛЕНИЯ (remote control); ВК (VK); СДС (SDS); АПД (ADF); КСС-СН (KSS-SN); ЦРРС (CRRS); КВ-УКВ Р/СТ (HF-VHF R/ST); КУЗОВ-КОНТЕЙНР СТАЦИОНАРНЫЙ (body container stationary); ПЕРЕВОЗИМЫЙ (carried); МОБИПЬНЫЙ (mobile). Mars

The operations of the Bal-E complex can be conducted 24 hours a day in fair and adverse weather conditions and in an NBC contaminated environment. The Kh-35E missiles can be launched when hidden behind natural terrain or man-made obstructions. Although referred to as a coastal system, Bal-E is can be positioned considerably inland, up to 10 km from the coast, and can be launched from below sea level to elevations up to 1000 m above sea level. From all positions the missiles can negotiate natural or man-made objects in their flight path. (Morinformsystem-Agat).

Among the required traits of the Bal-E complex was the need to be mobile with a very fast deployment/redeployment time. Each Bal-E unit is capable of relocating to a new fire position rapidly and coming into operation through means of high mobility and quick preparation for firing once at the new location. The coastal missile systems employ night vision capability, advanced navigation and positioning systems to aid ease of redeployment from one fire position to another in fair or adverse weather conditions day or night (Typhoon). Once in the new fire position the complex can be operational in around 10 minutes. In a high threat environment the ability to quickly relocate to a mew fire position would complicate targeting for counter strike forces as the unit could deploy to a fire position, fire a number of or the full complement of missiles and relocate in a very short period of time before a counter strike – missile or aircraft – could be accomplished. Relocation could be hastened by postponing missile reloading until the unit is redeployed to the new fire position, although this would, of course, increase the time required to reach missile launch readiness once at the new fire position. The self-propelled mobility of the Monolit-B complex allows it to change operational location very quickly in conditions of fair or adverse weather day or night when not operating as an integral part of the Bal-E coastal missile system (Typhoon). This, as with the Bal-E complex, reduces the probability of an effective counterstrike.

The high accuracy of the Kh-35E missile armament is courtesy of the integral ARGS-35E active-radar homing head contained in the missile nose section. Radar MMS

The Kh-35E forms the missile armament of the Bal-E complex. Prior to being selected as the missile armament of the ground launched Bal-E the Kh-35E had been adopted as a universal system for launch from tactical combat aircraft and surface warships. A Kh-35E is shown on an intermediate wing station of an RAC-MiG MiG-29 tactical strike fighter aircraft (top) and can be carried by other tactical combat aircraft of the Sukhoi Su-30 and Su-35S families. The Kh-35E constitutes the missile armament of the Uran-E ship mounted anti-ship cruise missile complex arming warships of the Project 12418 missile boats (above). Radar MMS

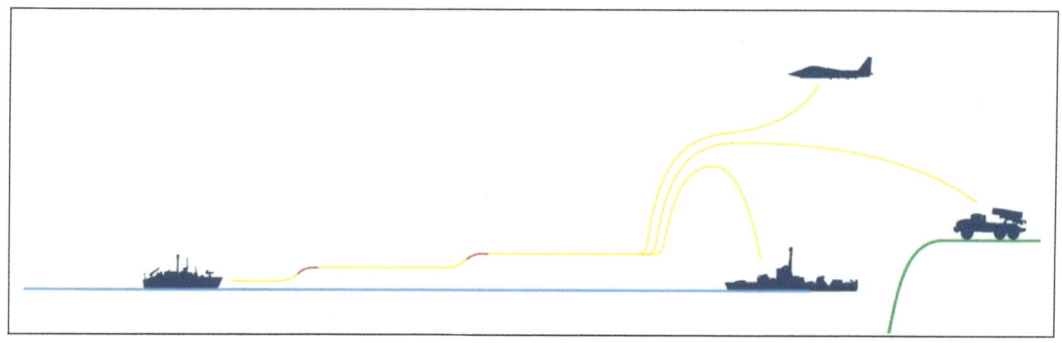

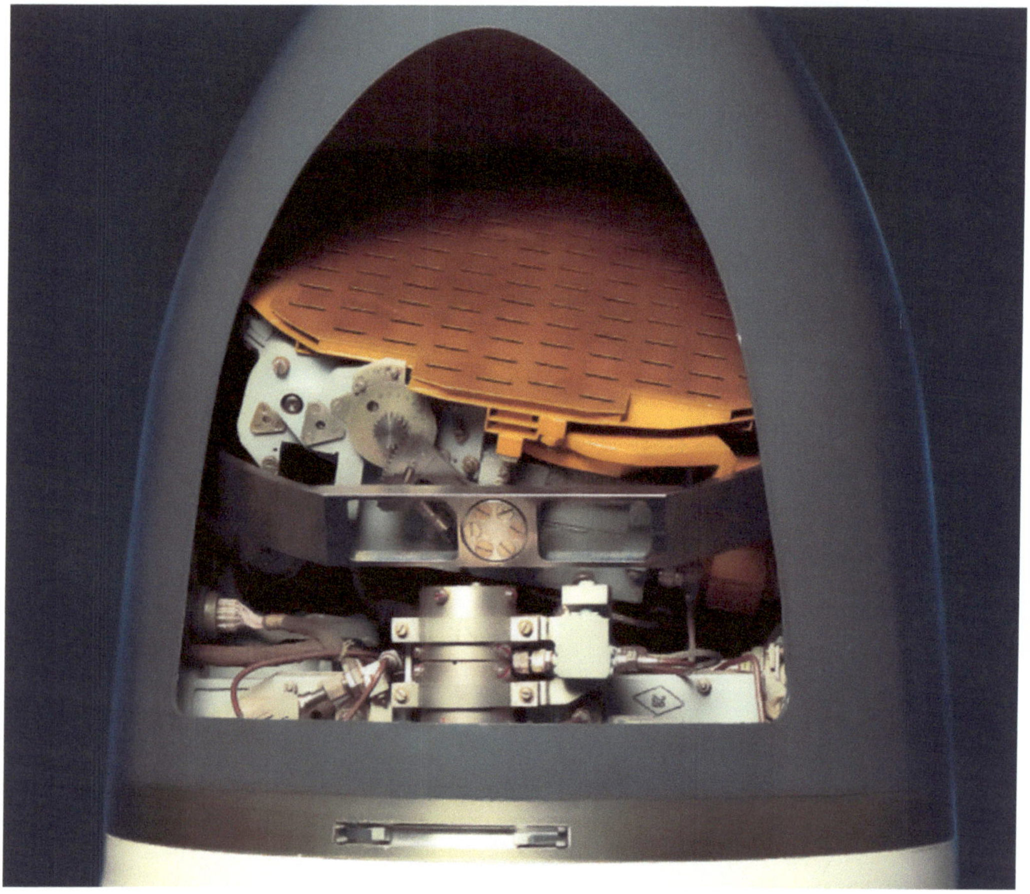

Top: Graphic depicting the launch of Kh-35E/UE anti-ship cruise missiles from the trio of available launch options – air launch from tactical combat aircraft, surface launch from Project 12418 missile boats armed with the Uran-E complex or surface launch from the land based mobile Bal-E complex. **Above:** Ghosted view of the Kh-35E forward section showing the antenna for the ARGS-35E active radar homing head. Radar MMS

Kh-35E (3M-24E) – The strike element of the Bal-E complex is the Tactical Missiles Corporation Kh-35E (3M-24E) ASCM, which is designed to destroy surface vessels, including warships displacing up to 5,000 tonnes. The missile, which can be air launched, launched from warships (Uran-E ship-borne missile system) and launched from Bal-E coastal missile batteries, is guided by an ARGS-35E active radar homing head. The ARGS-35E, which is employed in the terminal phase of the missile flight, was developed by the modern day Radar MMS. The ARGS-35E weighs less than 40 kg with basic dimensions of 420 mm diameter at its widest and a length of 700 mm (Radar MMS). The homing head allows timely detection of surface targets such as a warship underway and prioritisation of a single target when operating in a group. Once the target has been selected its position in both azimuth and elevation – the homing head provides accurate guidance in an azimuth range of ± 45° and an elevation range of + 10° to -20° – range and closing velocity of the target and the missile is determined. The target coordinates are updated and locked into the missile guidance system and the active homing head can detect targets at ranges out to 20 km in environmental conditions of an ambient temperature range of ± 50° C in precipitation conditions = to 8 mm/hr. day or night at sea states up to 6 (Radar MMS).

Described as immune to jamming (Morinformsystem-Agat), it is certainly the case that the ARGS-35E, which can employ PKR active noise homing when the missile is in the terminal phase of the trajectory (AGAT), has very high resistance to jamming in a dense countermeasures environment.

A Kh-35E ASCAM launched from a 3S-60E launch vehicle. Rosoboronexport

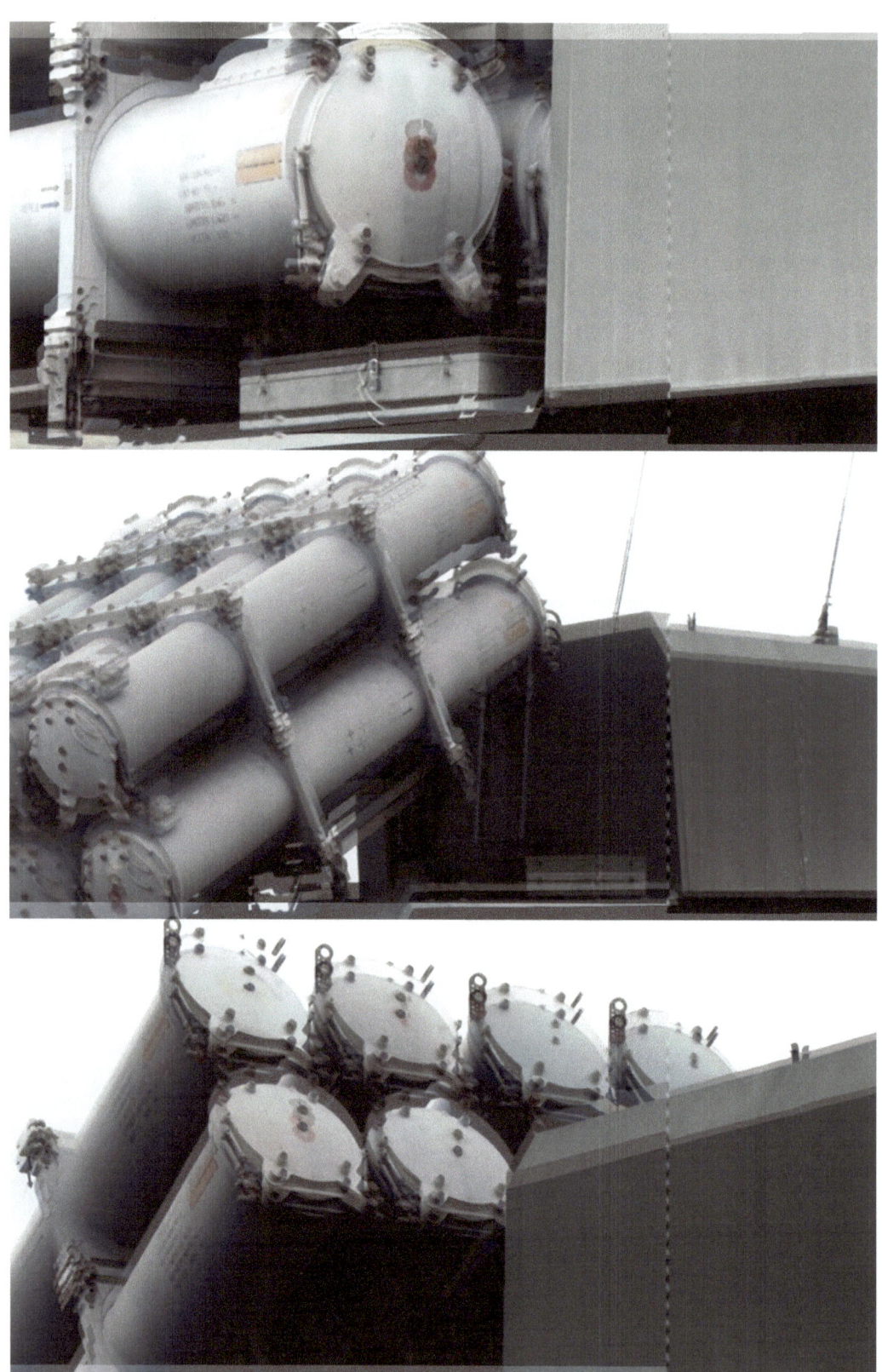

Previous page: Kh-35E missile launch containers of a Baltic Fleet Bal-E launch vehicle. Above: Kh-35E launch from a Bal-E launch vehicle: MODRF/Rosoboronexport

Once launched and stabilised, with maximum turn angle in horizontal plane after launch of ± 90°, the Kh-35E, which cruises at Mach 0.8, descends to an altitude of some 10-15 m above the sea surface, dropping to 4 m for the terminal phase of the flight, to strike targets up to 130 km distant in sea states up to 6 in an active electronic countermeasures environment. The ARGS-35E active radar seeker has an acquisition range of around 20 km, thereafter the target is locked-on and destroyed or disabled by the 145 kg high explosive penetrator warhead.

The Kh-35UE improves on the Kh-35E in a number of areas, including range, which is doubled from 130 km to 260 km, and features an improved post-launch horizontal turn capability of 130°. The Kh-35UE employs what is referred to as a combined guidance system that incorporates an inertial system with satellite navigation and an active/passive radar seeker head, which combines to increase accuracy. The Kh-35UE also features increased immunity to jamming measures/ECM (Electronic Counter Measures). The improved seeker head has a capability to lock-on to the target at a ranges out to 50 km, a considerable improvement on the 20 km lock-on capability of the Kh-35E seeker-head.

Kh-35E characteristics – data furnished by Tactical Missiles Corporation and Rosoboronexport

Launch range: up to 130 km
Missile flight altitude: 10-15 m en-route to the target area dropping to ~4 m in the terminal phase of the flight
Missile cruise speed: Mach 0.8
Maximum missile turn angle in horizontal plane after launch: +/- 90°
Launch weight: 620 kg (land based variant)
Warhead: 145 kg high explosive penetrator
Length: land launched variant, 4.4 m
Body diameter: 0.42 m
Wing span: 1.33 m

Kh-35UE characteristics – data furnished by Tactical Missiles Corporation and Rosoboronexport

Launch range envelope: 7-260 km
Missile flight altitude over wave ridge: 10-15 m en-route to target area dropping to ~4 m in the terminal phase of the flight
Missile cruise speed: Mach 0.8 to 0.85
Missile turn angle in horizontal plane post launch: +/- 130°
Guidance system: inertial plus satellite navigation plus active-passive radio homing
Maximum range of passive detection and lock-on with active-passive radio homing head: 50 km
Launch weight: 670 kg for surface launched variant
Warhead: 145 kg penetrating high explosive fragmentation
Length: surface launched variant, 4.4 m
Body diameter: 0.42 m
Wing span: 1.33 m
Weather conditions for use: any sea conditions up to sea state 6

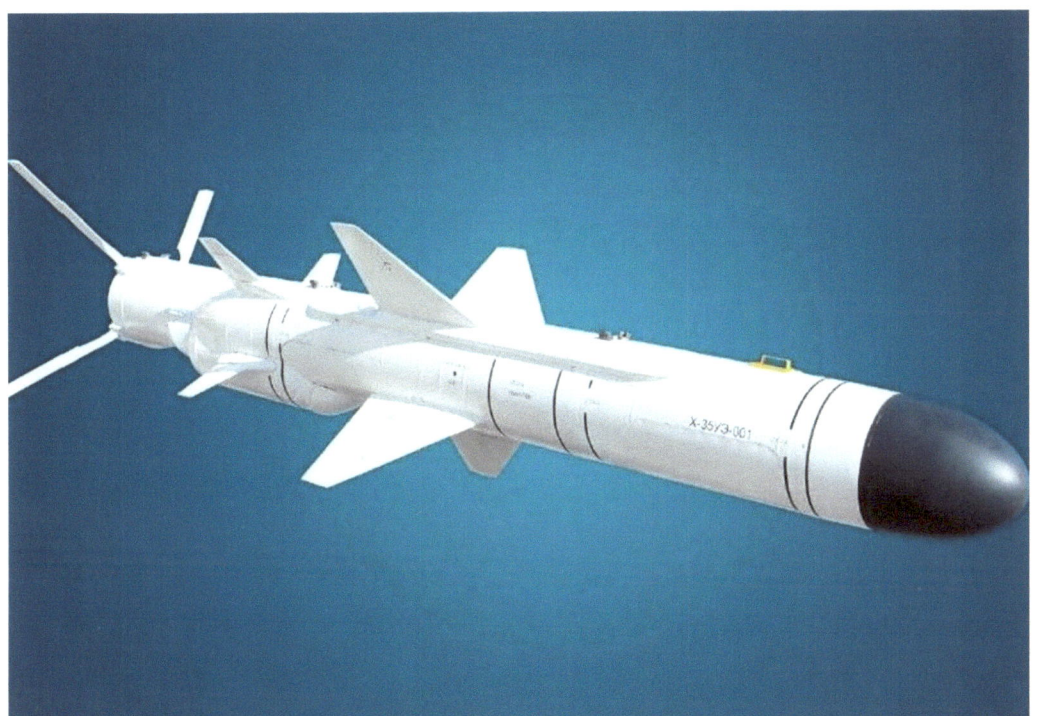

Kh-35UE (top) and Kh-35E launch containers (each LV has four missile containers) on a 3S-60E Bal-E LV (above). Rosoboronexport/MODRF

Following missile launch the guidance of the Kh-35E/UE can be conducted fully autonomously 24 hours a day in environmental conditions of fair or adverse weather in a dense missile/gun hard kill and electronic counter measures defensive environment (TMC). The complex receives radar tracking data, including the various parametric information for each tracked target.

Bal-E complex of the Russian Federation Baltic Fleet – launch vehicle (top) and Monolit-B command vehicles (above). MODRF

Russia's Coastal Missile Shield

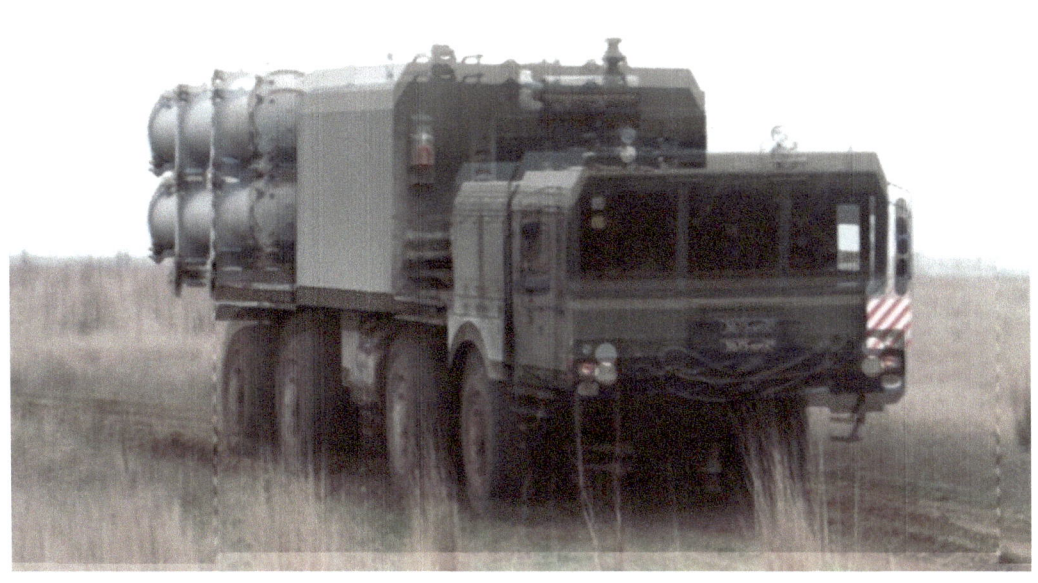

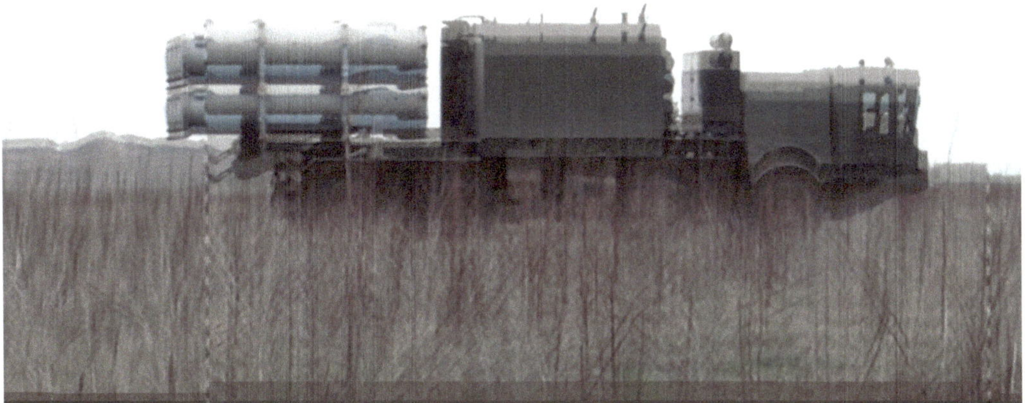

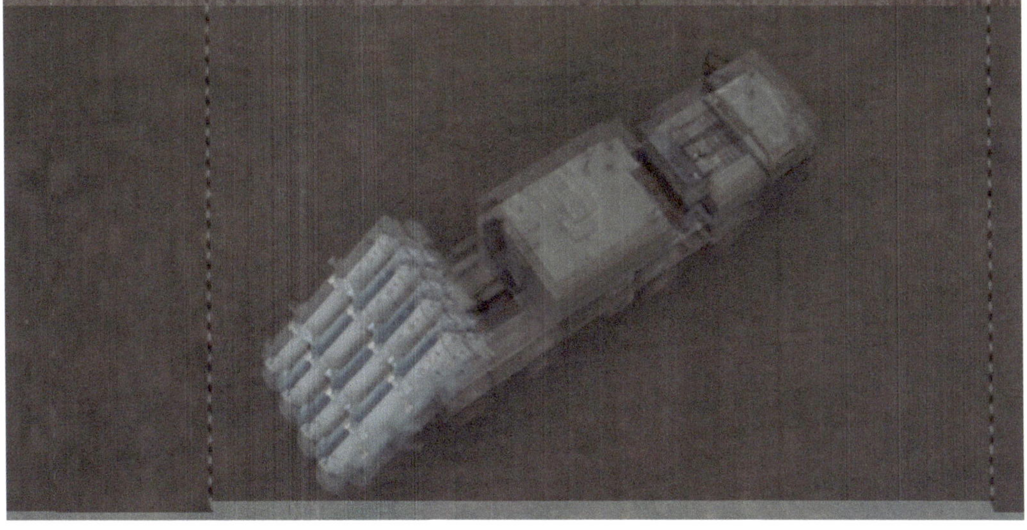

27

Page 27-28: Bal-E complex self-propelled launch vehicles of the coastal forces of the Russian Black Sea Fleet. MODRF

3

BASTION

The Onyx missiles complex – sea based – armed with 3M-80E Yakhont supersonic cruise missiles were developed in the late Cold Wear period. The first platform for the Onyx system was the Project 1234.7 small missile boats of the Nakat type, which were armed with twelve Yakhont missiles from 1991 (Mashinostroyenia). The Yakhont cruise missile also constitutes the missile element of the Bastion CMS (Coastal Missile System), which entered service with the coastal forces of the Russian Federation in 2015 (MODRF & Mashinostroyenia).

The Bastion complex consists of a number of elements that include the self-propelled TLV (Transport Launch Vehicle) – each with two Yakhont missiles carried in TLC (Transport Launch Containers); mobile command post; light crew support vehicle, reloading vehicle carrying additional missiles in TLC, various equipment sets for data collection, interfaced with the missile armament and the necessary maintenance and training aids (Mashinostroyenia). The TLV and the reloading vehicles are based on the same MZKT-7930 chassis as adopted for the Bal-E complex (Mashinostroyenia & Volat). A typical Bastion battery consists of 18 self-propelled TLV; transport reloading vehicles with machinery for loading of missiles in TLC; three command post vehicles fitted out with automated control systems and a communications suite; alert support vehicles (number unspecified); maintenance support systems (presumably vehicle mounted) and 38 ready to launch Yakhont supersonic anti-ship/land attack cruise missiles in TLC – two mounted on each TLV (38 missiles would be carried in the transport reload vehicles. The Yakhont missile can engage all types of maritime surface as well as radio-contrast land targets. The Mach 2.5 flight speed, a range of flight profile options, the missiles low radar cross section and an advanced fully-autonomous, stated 'jam-resistant', guidance system that incorporates a modern inertial navigation system and an active-radar seeker head make the Yakhont very difficult to defend against. Once the onboard seeker acquires the target the missile is automatically guided (Rosoboronexport).

The Yakhont shipborne automatic control system (equips the land based Bastion complex), which facilitates the missiles ability to fly at very low-altitudes, and the homing head (HH-Yakhont), designed as a modular system – antenna, transmitter, receiver and processing unit – was developed by JSC Concern Granit-Electron. The two-channel system, which is capable of operating in a dense EW (Electronic Warfare)/jamming and counter fire environment employs passive and active channels in search and tracking modes. In the active mode the system uses a wideband coherent signal (Granit-Electron). The homing head is optimised to detect maritime surface and land targets and can operate in a dense ECM (Electronic Countermeasures) environment. The system is designed to operate against active and passive countermeasures in a high jamming environment to eliminate false targets. Once a target is confirmed and selected the relevant coordinates are relayed to the missile auto-pilot in conditions of fair or adverse weather with sea states up to 7 (Granit-Electron).

Bastion complex TLV with the missile launch containers in the ready to launch position. Mashinostroyenia

Page 32-33: Bastion TLV's of the Russian Federation Northern Fleet. MODRF

The Bastion complex Command Vehicle is a Kamaz 6x6 unit, apparently based on the Kamaz 43101 while the light crew support vehicles is a Kamaz unit also (model unconfirmed at the time of publication). Mashinostroyenia /MODRF

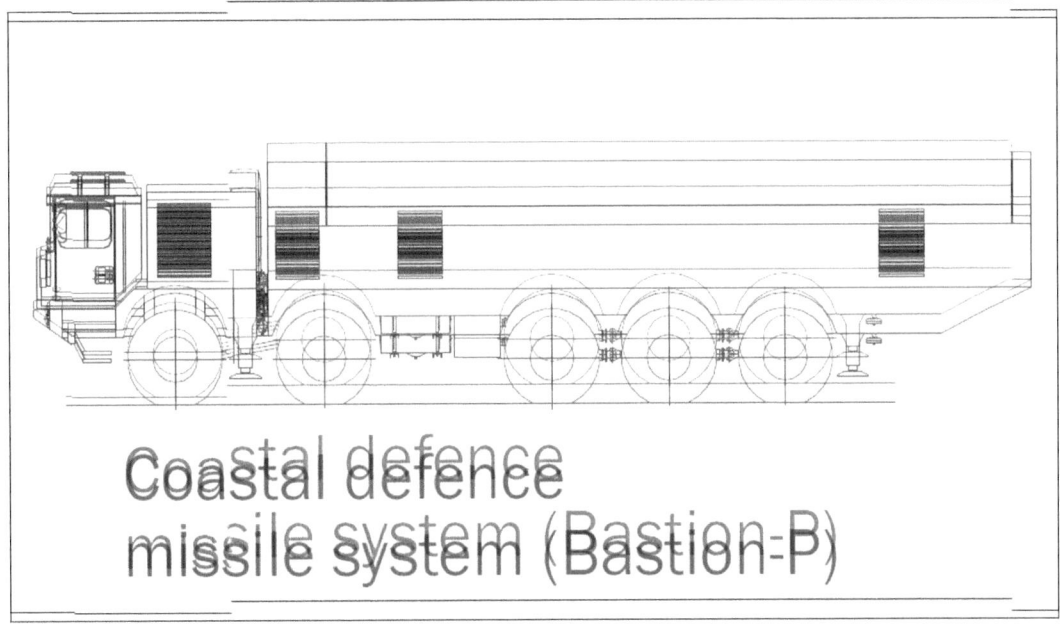

Top: An operational Bastion unit during deployment transit in Russia. Above: While the Bastion complex in service with the Russian Federation centred on the MZKT-7930 vehicle an extended wheel variant of the MZKT-6002 family is offered as an option for a complex referred to as Bastion-P. MODRF/Volat

The Yakhont missile of the Onyx complex was developed in the late Cold War period and first deployed on a surface combatant in 1991, the year the Soviet Union was dissolved to form a Commonwealth of Independent States. The basic aerodynamic layout included a nose mounted air intake with a set of cruciform main wing surfaces at roughly two thirds from the missile fore with a set of smaller wing surfaces just behind. *Avangaurd/Strela*

Top: The first platform to be armed with the 120 km range 3M-80E (range for the 3M-80E1 is 110 km) Yakhont missile was the Project 1234.7 small missile boats of the Nakat type. These vessels, the first of which was commissioned into the Soviet Navy in 1991, were armed with twelve such weapons. The Project 1234 were transferred to the Russian Federation Fleets following the break-up of the Soviet Union. Above: The Yakhont missiles of the Bastion coastal missile system are housed in self-contained launch containers, although these differ in detail from those of the ship launched variant. Rosoboronexport

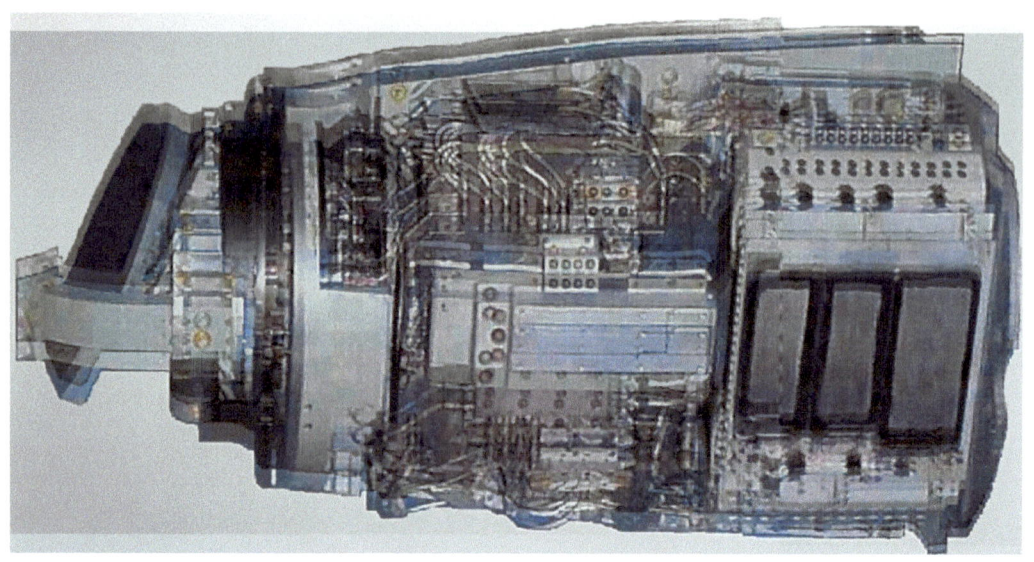

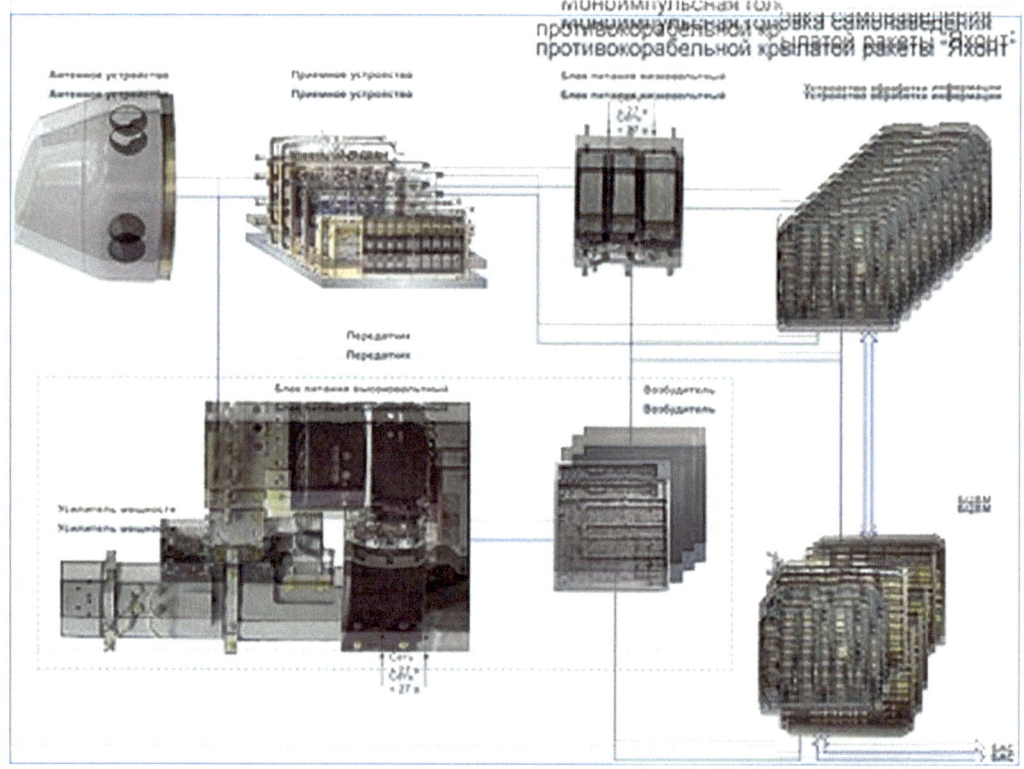

HH-Yakhont characteristics – data furnished by JSC Concern Granit-Electron
Target detection range in active mode: ≥50 km **Maximum target search angle:** ± 40° **Operational readiness time once switched on:** ≤2 minutes **Current consumption ±27 V:** ≤30A **Weight:** 75 kg

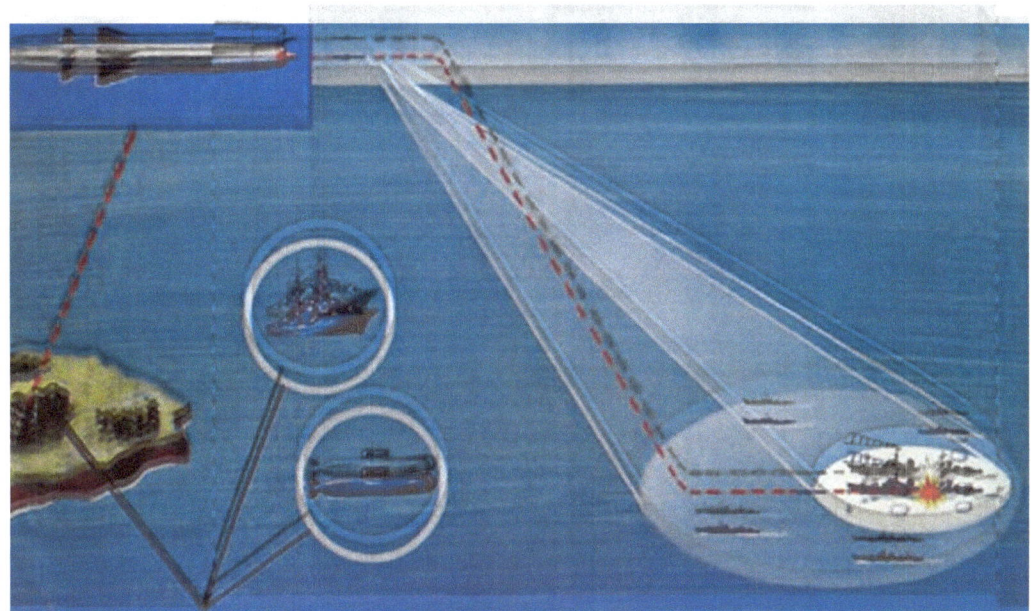

Previous page top: The Yakhont missile seeker-head, referred to as the 'HH-Yakhont'. Previous page bottom: Diagram showing a break-down of the various elements that make up the 'HH-Yakhont' seeker-head. This page top: Diagram showing the targeting process for a Bastion missile system. Data regarding a threat naval group approaching Russian shores is transferred to the Bastion complex by surface and sub-surface assets (data can be transferred by other assets such as aircraft or orbiting satellites), the missile is launched and follows a high altitude flight trajectory – up to 14000 m – in the first phase of the flight before dropping to ultra-low-altitude (around 10-15 m above the sea surface) in the terminal phase of the flight. Above: Development launch of a Yakhont. Granit Electron/Mashinostroyenia

A Yakhont missile is launched from a Bastion launch vehicle of the Russian Federation Northern Fleet located on the island of Boiler in Russia's Arctic Far East. MODRF

The Yakhont missile wings and fins are folded when stored in the sealed TLC from which the missile does not have to be removed during routine checks (Mashinostroyenia). When launched in the vertical attitude the missile reaches an altitude of ~60-80 metres, at which point pyrotechnics arrest the vertical momentum through a serious of burns – a burn in the forward cap pushes the missile toward the horizontal attitude, another, at the opposite end of the forward cap, arrests the horizontal push before two simultaneous burns propel the cap away from the missile, which continuous on its flight profile in the more or less horizontal attitude. The forward cap that was propelled away from the missile falls back to the surface a short distance, perhaps 100-200 m, or so from the launch platform.

For a maximum range of a combined hi-low flight profile of 300 km at a mean 720 m/s total flight time would be around 5 minutes 53 seconds. At a maximum range of 120 km for a low altitude profile the missile flight time would be 2 minutes 21 seconds. A typical subsonic ASCM would have a flight time of roughly 2.7 times longer, which would equate to 5.975 minutes for a low altitude flight profile. The vastly reduced flight time over that of a subsonic ASCM flying the same flight range profile increases missile survivability by reducing the available time for shipborne or shore based defenses to react, which may be as little as three or four seconds from detection to impact on the ship/land target. Even modern shipborne air defence systems of the MBDA Aster 15-30- series (Aster 15 maximum velocity is Mach 3 whilst Aster 30 velocity is Mach 4.5 (MBDA) (it should be noted that the maximum velocity of potential targets for successful interception would be below these velocities, but no values have been forthcoming from MBDA) would be extremely hard pressed to combat a Yakhont missile flying at Mach 2.5 10 metres above the waves. With the very short warning time from detection to impact Yakhont possess a considerable threat to any surface target due to the extremely low reaction time available for the defences to launch, acquire and attempt an interception.

Yakhont missile characteristics – data furnished by JSC MIC Mashinostroyenia, PA Strela & Rosoboronexport

Maximum flight range: 300 km when flying a combined hi-low flight trajectory and 120 km when flying a 100% low trajectory
Maximum distance from MAL [launch vehicle] to the coast: up to 200 km (JSC MIC Mashinostroyenia) 250 km (Rosoboronexport)
Flight altitude: up to 14000 m dropping to 10-15 m in the terminal phase
Maximum flight velocity: up to 750 m/s (Mach 2.5) in the cruise phase of the combined trajectory flight and 680 m/s in the terminal phase of the flight
Missile weight: 3000 kg (JSC MIC Mashinostroyenia & Rosoboronexport) and 3100 kg (PA Strela). For contrast the 3M-80E ship launched variant weighs 4150 kg and the 3M-80E1 weighs 3970 kg
Missile weight when housed in TLC: 3900 kg
TLC dimensions
    length: 8900 mm
    diameter: 720 mm
Warhead weight: 200 kg (the 3M-80E/E1) ship launched variant has a 300 kg penetrator warhead
Flight control system: on-board automatic control system/radar homing – inertial navigation system/radar altimeter/radar homing seeker
Propulsion: straight flow-Aerojet
Booster stage: solid fuel propellant
Sustainer type: ramjet
Sustainer fuel: kerosene T-6
Missile launch interval: 2-5 seconds
Launcher inclination: 15°-90°
Launch readiness time from power off: no more than 4 minutes (Rosoboronexport). For MAL and MCP [mobile command post] up to 5 minutes (JSC MIC Mashinostroyenia)
Period of independent duty: 5 days
Radar seeker target detection range: 75 km
Time between missile in TLC scheduled inspections: 3 years
Specified lifetime: 10 years

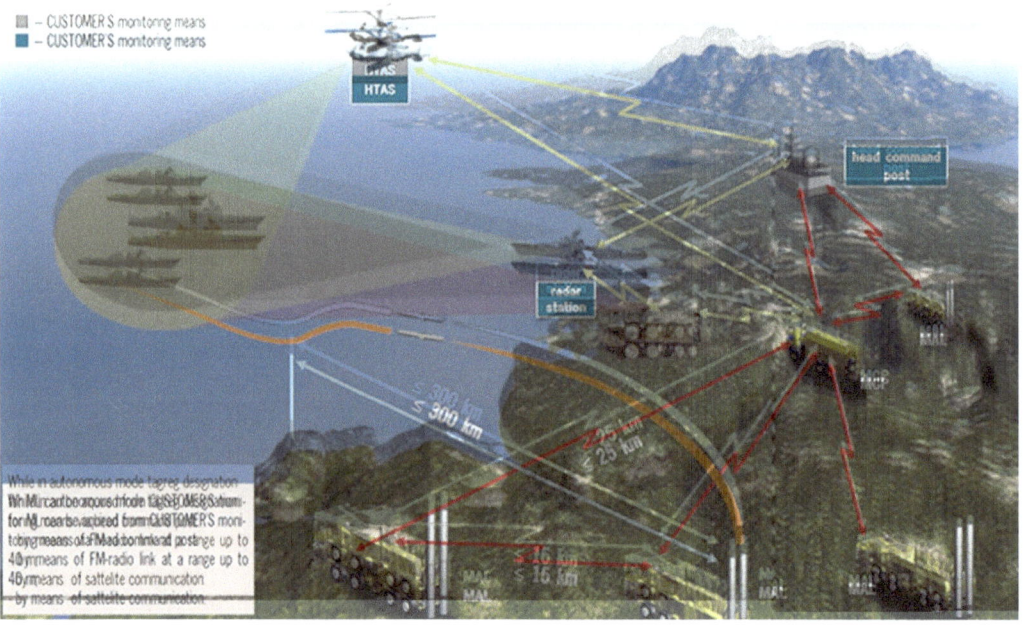

Graphic depicting the operation of a mobile Bastion coastal missile complex: Mashinostroyenia

Page 43-44: Series of stills showing the launch of a Yakhont missile from a Bastion complex of the Russian Federation Northern Fleet. A pyrotechnic burn pushes the missile to the near horizontal attitude, another burn arrests the horizontal push before two simultaneous burns propel the cap away from the missile, which continues on its cruise flight. MODRF

Page 45-46: Bastion complex – Russian Baltic Fleet based in Kaliningrad. MODRF

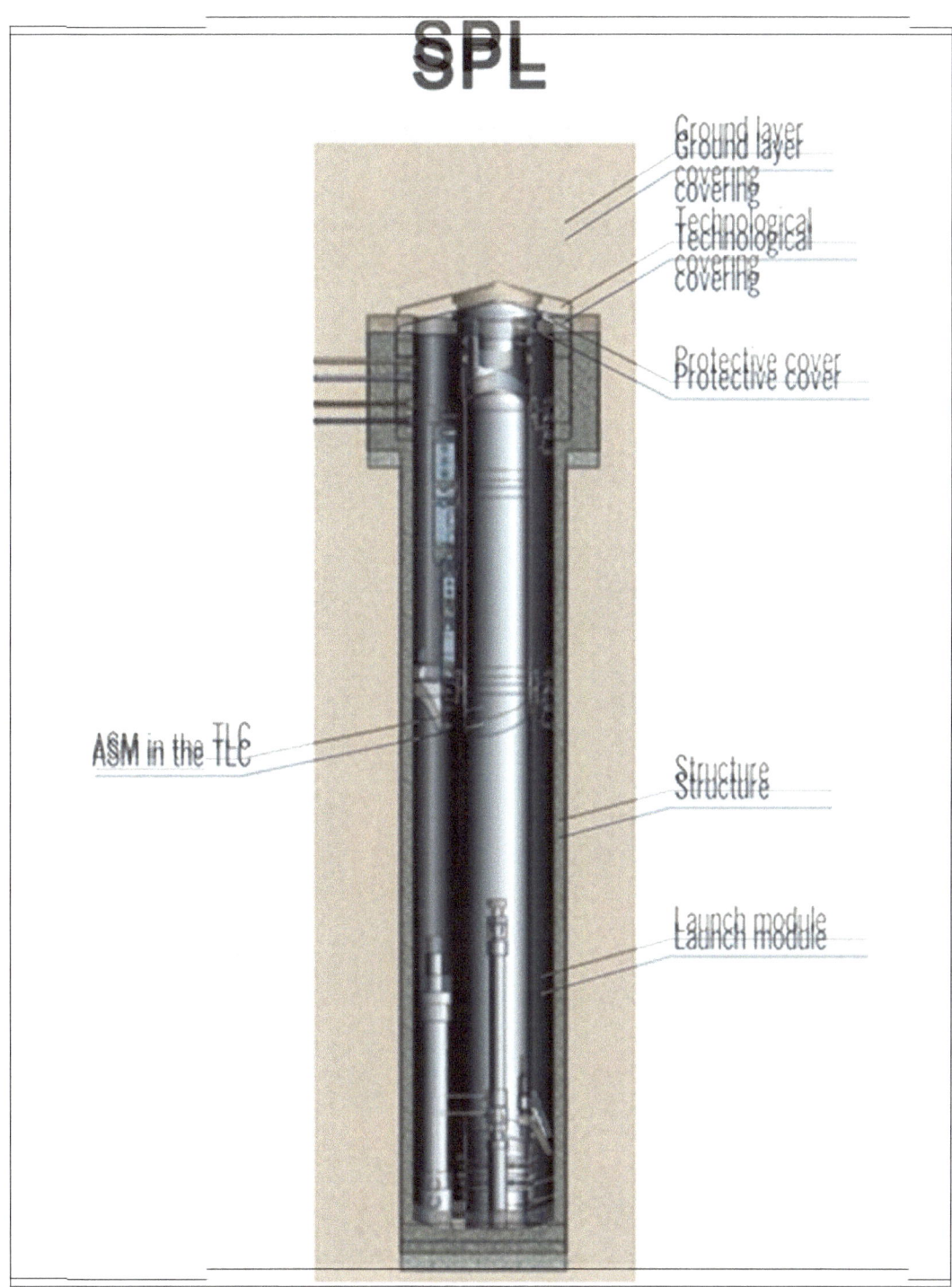

Diagram of the a stationary Bastion coastal missile system missile silo showing the various parts of the complex – ground layer covering; technological covering; protective cover; silo internal structure; Yakhont missile in the transport launch container and the launch module. Mashinostroyenia

A test launch of a Yakhont missile from a stationary Bastion coastal missile system silo structure. Mashinostroyenia

Bastion can also be deployed in stationary shore based form with the same overall engagement capabilities as the mobile shore based system – can cover 600 km length of shore covering a total sea area of 150000 km^2. The stationary system would be suitable for deployment in protection of naval bases, ports and strategic coastline where enemy assault landings would be expected, but would be more vulnerable to attack from aircraft or missile assets and, therefore, require additional protective measures compared to that required by mobile system. While camouflaging of the various sites associated with the stationary Bastion system would constitute a line of defence, the various elements would be concealed in underground facilities including the missiles, which would be housed in SPL (Stationary Protected Launchers) in special protective launch silos. The stationary Bastion complex incorporates a missile battery in the SPL's and associated command and control modules. As is the case with the mobile complex the missiles of the stationary complex are housed in sealed TLC with the wings and fins in the folded position. The SPL consists of the launch module; silo structure; Yakhont missile in the TLC; protective upper cover; technological covering and the ground layer (Mashinostroyenia).

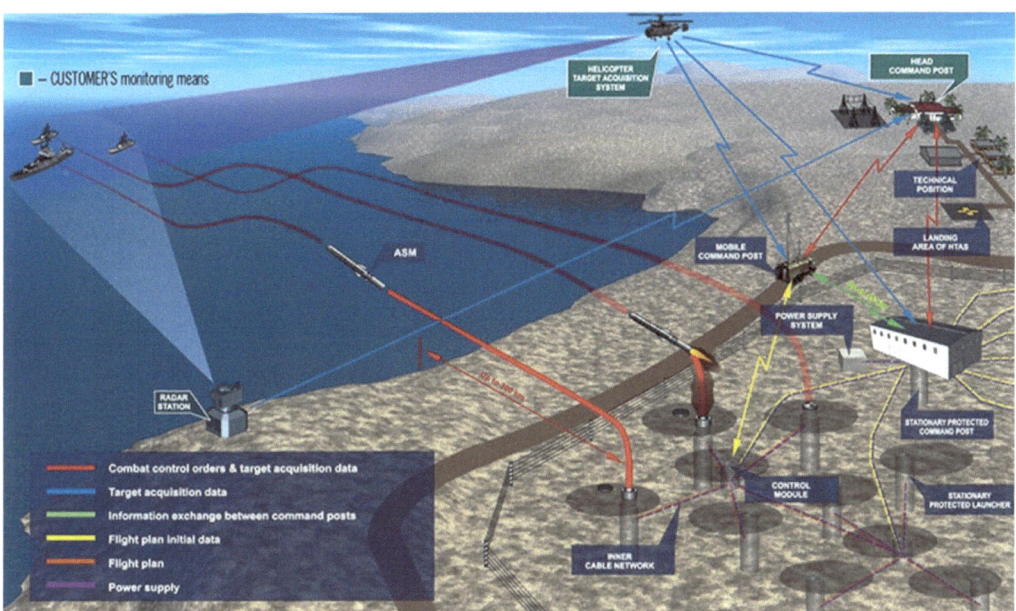

Graphic depicting the operation of a stationary Bastion complex. Mashinostroyenia

The primary threat Bastion and Bal-E are designed to counter is enemy landing forces, including warship escorts. However, the 300 km range for Bastion and 130 km range for Bal-E (Kh-35E missile) would push NATO (North Atlantic Treaty Organisation) cruise missile carriers and aircraft carrier battle groups farther away from Russian Federation shores, reducing the range of targets that could be engaged with BGM-109 Tomahawk LACM (Land Attack Cruise Missiles). Longer flight time of the LACM would increase their vulnerability to defensive EW countermeasures and the warning time for hard kill defensive systems like S-400 and Pantsir surface to air missile systems. In the event of unavailability of other longer range systems such as air launched Kh-32 high supersonic (potentially low hypersonic – maximum velocity value has not been released as of late summer 2018), air launched Kh-47 hypersonic and subsonic 3M-54E Kalibr-NK ASCM launched from surface warships and submarines, Bastion and Bal-E would form a defensive missile screen against hostile battle groups approaching Russian shores to increase the range of their tactical combat aircraft or in support of an assault landing operation, which would also be a priority target set for Bal-E and Bastion.

Bastion launch vehicle of the Russian Black Sea Fleet (top) and a missile launch form a Bastion launch vehicle of the Northern Fleet (above). MODRF

4

OPERATIONAL

Bal-E and Bastion coastal missile systems are in service with the coastal forces of the major fleets of the Russian Federation. Bal-E entered service with the Pacific Fleet Missile Brigade in late 2014 and conducted missile firings in the Primorye Territory in April the following year. Bal-E was operating with the Separate Missile Brigade of the Black Sea Fleet in 2015 and by early 2018 was established with the Black Sea Fleet missile formation in Crimea (MODRF). The complex is also in service with Russian Baltic Fleet coastal forces, including those based in Kaliningrad, Russia's forward bastion in Eastern Europe, which is surrounded by NATO (North Atlantic Treaty Organisation) nations, other than the Baltic Sea coastline. Bastion, which entered service in 2015, was allocated to the Northern and Black Sea Fleets prior to entering service with the Pacific Fleet Missile Artillery Brigade (Mashinostroyenia), which, at the start of April 2016, began receiving Bastion complexes in the Primorsky Krai, to cover the Vladivostok region of the Russian Far East (MODRF). Bastion is also in service with the coastal forces of the Baltic Fleet, including a battery based in Kaliningrad. Kaliningrad, which has the only Russian Baltic coastline that is expected to be ice free all year round, is important to the Russian defence against the increasing NATO military presence on or near her borders.

Bastion has increased range, increased survivability and increased lethality compared with Bal-E through its high supersonic cruise speed and more powerful warhead. Bal-E has the advantage of higher missile load out – up to eight per launch vehicle. This allows saturation attacks to be conducted against high value threat targets as well as the ability to better counter amphibious assault landing forces that would entail targeting large numbers of surface targets. Of course, both systems are complementary to each other, providing multiple coastal missile system shore based layers of the overall anti-ship missile protection for the Russian Federation coasts and extended protection into the various seas and oceans afforded by the layered defence of air, submarine, surface ship and land launched anti-ship missiles. The inner zone of the Russian Federation coast defence remains the Bereg-E self-propelled coastal artillery complex.

Top: Bal-E self-propelled launch vehicle of the Baltic Fleet Missile Brigade located in Kaliningrad. Above: Bal-E launch vehicle of the Black Sea coastal Missile Brigade. MODRF

Bal-E launch vehicles. MODRF

Top: Kh-35E missile launched from a Bal-E launch vehicle of the Russian Black Sea Fleet. Above: Bal-E launch vehicles of the Russian Black Sea Fleet. MODRF

Previous page top: Monolit-B complex of the Black Sea Fleet with a Bastion complex launch vehicle in the background. Previous page bottom: Monolit-B and Bastion complex launch vehicles of the Russian Pacific Fleet. Above: Bastion self-propelled launch vehicle of the Russian Pacific Fleet. MODRF

Bastion was used operationally in the land attack role in 2016 when a handful of Yakhont missiles were launched and successfully struck extremist targets in the Syrian Arab Republic in support of Russian/Syrian Arab Army operations against Islamic state and other groups attempting exert control over Syria (MODRF).

Bal-E and Bastion constitute two links in the chain of coastal missile defence in depth currently practiced by the Russian Federation. This includes surface vessels armed with several types of anti-ship cruise missiles complexes (including Uran-E armed with the Kh-35E), submarine and surface ship launched anti-ship cruise missiles of the P-700 Granite (supersonic) complex and submarine and surface ship launched Kalibr-NK (subsonic) cruise missiles, long range torpedoes, aircraft launched anti-ship cruise missiles – Kh-22 (high supersonic), Kh-32 (high supersonic (possibly low hypersonic)) and Kh-47M (designation unconfirmed in 2018) Kinzhal (hypersonic) – and older generation costal launched anti-ship cruise missiles – some units continue to operate Redut and Rubezh in 2018, but these complexes are increasingly being phased out of service.

The Club-M complex is not in service with the costal forces of the Russian Federation, but the system is offered for export as an alternative to Bastion and Bal-E. Club-M is a land based variant of the Club-S (submarine) and Club-N (surface warship launched complexes utilising the 3M-54E and 3M-54E1 anti-ship missiles and 3M-14E land attack cruise missile). Like Bastion, Club-M is designed for striking surface vessels and land targets, but at longer range and reduced flight speed.

In 2018, the Redut (top) and Rubezh (above) missiles complex remain in service with the coastal forces of the Russian Federation, but are expected to be completely phased out over the next several years. MODRF

The Bereg (Beach) self-propelled coastal artillery complex constitutes the inner element of the Russian Federation layered coastal missile/gun defence. Bereg can engage targets out to ranges of 23 km. While there is, in 2018, no replacement for the Bereg the Bal-E (above) and Bastion are establishing themselves in service as Redut and Rubezh are phased out of service. Titan/MODRF

GLOSSARY

ARdr	Active Radar Channel
ASCM	Anti-Ship Cruise Missile
bit/s	Bits per second
C3	Command Control & Communications
CCC	Command Control & Communications
CDB	Central Design Bureau
CMOP	Coastal Modular Operational Post
CMS	Coastal Missile System
DBM	Dubna Machine Building Plant
ECM	Electronic Countermeasures
EW	Electronic Warfare
HF	High Frequency
hp	Horsepower
IFF	Identification Friend or Foe
JSC	Joint Stock Company
kbit/s	Kilobits per second
Kg	Kilogram
kg/cm	Kilogram per centimetre
Km	Kilometre
km²	Kilometre squared
Km/h	Kilometre per hour
kW	Kilowatt
LACM	Land Attack Cruise Missile
LCD	Liquid Crystal Display
LV	Launch Vehicle
m	Metre
Mach	1 Mach – the speed of sound (this varies with altitude)
MiG	Mikoyan
mm	Millimetre
mm/hr	Millimeter per hour
MODRF	Ministry of Defence of the Russian Federation
m/s	Metres per second
NATO	North Atlantic Treaty Organisation
NBC	Nuclear Biological Chemical
PRdr	Passive Radar Channel
RAC	Russian Aircraft Corporation
TLC	Transport Launch Containers
TLV	Transport Launch Vehicle
TMC	Tactical Missiles Corporation
Tu	Tupolev
USSR	Union of Soviet Socialist Republics
x	Times multiplication

%	Percent
±	Plus or minus
≤	Less than, or equal to
≥	More than, or equal to
°	Degree(s)
~	Approximately equal to (can also be used to mean asymptotically equal)

ABOUT THE AUTHOR

Hugh Harkins FRAS, MIstP, MRAeS is a physicist/historian and author with an extensive research/study background in aeronautic, astronautic, astrophysics, geophysics, nautical and the wider scientific, technical and historical fields. He is also involved in research in the field of Scottish history, which formed a significant element of dual undergraduate degrees. Hugh has published in excess of sixty books, non-fiction and fiction, writing under his given name as well as utilising several pseudonyms. He has also written for several international magazines, whilst his work has been used as reference for many other projects, ranging from the aviation industry, international news corporations and film media to encyclopaedias, museum exhibits and the computer gaming industry. Hugh an elected member of the Institute of Physics and Royal Aeronautical Society and is an elected Fellow of the Royal Astronomical Society. He currently resides in his native Scotland. Other titles by the author include:

Iskander - Mobile Tactical Aero-Ballistic/Cruise Missile Complex
Orbital/Fractional Orbit Bombardment System - The Soviet Globalnaya Raketa
Counter-Space Defence Co-Orbital Satellite Fighter
Sukhoi T-50/PAK FA - Russia's 5th Generation 'Stealth' Fighter
Sukhoi Su-35S 'Flanker' E - Russia's 4++ Generation Super-Manoeuvrability Fighter
Sukhoi Su-34 'Fullback'
Sukhoi Su-30MKK/MK2/M2 - Russo Kitashiy Striker from Amur
MiG-35/D 'Fulcrum' F – Towards the Fifth Generation
Air War over Syria, Tu-160, Tu-95MS & Tu-22M3 - Cruise Missile and Bombing Strikes on Syria, November 2015-February 2016
Sukhoi Su-27SM(3)/SKM
Russian/Soviet Aircraft Carrier & Carrier Aviation Design & Evolution Volume 1 - Seaplane Carriers, Project 71/72, Graf Zeppelin, Project 1123 ASW Cruiser & Project 1143-1143.4
Heavy Aircraft Carrying Cruiser
Light Battle Cruisers and the Second Battle of Heligoland Bight
British Battlecruisers of World War 1 - Operational Log, July 1914-June 1915
Eurofighter Typhoon - Storm over Europe
Tornado F.2/F.3 Air Defence Variant
Air to Air Missile Directory
North American F-108 Rapier - Mach 3 Interceptor
Convair YB-60 - Fort Worth Overcast
Boeing X-36 Tailless Agility Flight Research Aircraft
X-32 - The Boeing Joint Strike Fighter
X-35 - Progenitor to the F-35 Lightning II
X-45 Uninhabited Combat Air Vehicle
Into The Cauldron - The Lancaster MK.I Daylight Raid on Augsburg
Hurricane IIB Combat Log - 151 Wing RAF, North Russia 1941
RAF Meteor Jet Fighters in World War II, an Operational Log
Typhoon IA/B Combat Log - Operation Jubilee, August 1942
Defiant MK.I Combat Log - Fighter Command, May-September 1940
Blenheim MK.IF Combat Log - Fighter Command Day Fighter Sweeps/Night Interceptions, September 1939 - June 1940
Tomahawk I/II Combat Log - European Theatre, 1941-42
Fortress MK.I Combat Log - Bomber Command High Altitude Bombing Operations, July-September 1941
XF-92 - Convairs Arrow

www.ingramcontent.com/pod-product-compliance
Lightning Source LLC
Chambersburg PA
CBHW051917210526
45473CB00006B/2052